I would like t

____

____

____

____

____

Faithfully yours,

Sign.: ________________________

Name:________________________

Date: ________________________

Eng. Faris Rashid AlHajri
Website: www.farisalhajri.com
Email1: faris@farisalhajri.com
Email2: qaishigh1@yahoo.com

The Miracle & Wonders of Treatment from Hot Water

Hot Water Miracles

Faris Alhajri

authorHOUSE®

AuthorHouse™ UK Ltd.
500 Avebury Boulevard
Central Milton Keynes, MK9 2BE
www.authorhouse.co.uk
Phone: 08001974150

First published by AuthorHouse 9/27/2010

ISBN: 978-1-4490-9778-3 (sc)

"Wealth does not bring happiness,
but happiness brings wealth.
Neither Happiness, nor wealth
brings health,
But health brings both happiness
and wealth.
Every human's ambitions are to
possess health, happiness and
wealth.
But wealth does not come without
happiness
And happiness does not come
without health.
Every human whether baby, child,
youth or adult struggles for health
in order to survive.
The dream of every human is to live
a long life filled with health,
Until they accomplish the star that
has been a gift to them since birth
in the same manner of our longest
living ancestors".

"Every leader, a sultan, president, minister, engineer, politician, science experts and religious leaders are ambitious to achieve accomplishments on their own field. Both people in the local and international level have different objectives that they want to achieve.

A greatest accomplishment that is recorded on the history is those which maintain peace and unity of the people around the world. Among them is His Majesty Sultan Qaboos Bin Said, has managed to accomplish numerous recognition or awards on the field of Peace, Environment, Unity and Advocacy to Human Rights".

Inscription

This book is dedicated to all the People of the World. Any kind of discrimination whether in Religion, Doctrine, Resemblance, Color, Size, Age, Aspect, Financial Potency, Authority... etc, should be disregarded.

Due to the benevolent life that has been provided by my beautiful Country - the Sultanate of Oman - under the leadership of His Majesty- Sultan Qaboos Bin Said, every Citizen and Resident has been enjoying a clean Environment, Integrity and Stability. Proper education has been provided. However, all these facilities have been playing a big role for what I have achieved so far.

My special dedication is also passed to every reader of this book, mostly, my beautiful Wife (Gloria), and my lovely Children, (Qais & Sami). Including, everyone who knew me personally, and those who tendered their emotional supports, whereas I was spreading this scientific achievement.

Let us remember and strongly believe that:

"Health and happiness are the most prominent wealth that every human should possess.

People who possess extreme health and happiness are the one who exert large efforts to make others happy and healthy, by helping them whenever they discover the secret for achieving more health and happiness.

A quality wealth that human should possess must be shared to others for that contribution will remain forever and that wealth will be considered divine for it the product of love".

Acknowledgement

God, the Almighty said: "And whoever does, any, righteous deeds, whether male or female, and is a believer- such shall be admitted into Paradise, and not be wronged, by as much as, the didnt in a date-stone" from the Quranic Version, ALNISAA-((WOMEN))- (4:124)

What a wonderful World we live, where I am blessed to be surrounded by many glorious People in my entire life. Most especially, my beautiful Wife Gloria, my handsome Children, Qais & Sami, who have offered me a foundation of motivation and unforgettable support.

I should never forget the emotional supports I have been receiving from those leaders, who have shown an expression of modesty; whose names have not been revealed here.

I must also present my special thanks to my brother, Mr. Hilal Rashid S. AlHajri who always appreciated my works. Of course, I also thank my brother-in-law Mr. Hamood Khalfan S. AlBusaidi who has always been very conversational with words of knowledge and genius.

It is also important to tender my sincere gratefulness to all my friends and those who have contributed to the wonderful peaceful environment I live; Dr. Pratip Bhattachargee-MD-MRCP (UK), Mr. Hassan Mohammed A. AlAnsari, Eng. Abdelqawi Abdullah A. AlYafei, Eng. Abdullah Rashid S. AlKayumi, Mr. Hassan Ahmed H. AlShaqsi, Mr. Abdullah Hamad H. AlHarthi, Mr. Usama AlSaed Ateyah, Mr. Ahmed Jalal Mohammed, Ms. Jennifer Buenconsejo, Ms. Susanna Rose Georgena King, Mr. Rashid Said A. AlHabsi, Mr. Ghalib Said A. AlHarthy, Mr. Fuad Rashid S. AlHajri, Mr. Rashid Hamood M. AlToqi, Mrs. Fatma Ali S. AlHajri, Mr. Khalifa Ali S. AlHajri, Mrs. Maryam Salim S. AlToqi.

It is also a sign of proud for Ms. Maji Ceneta, Mr. Nasser Abdullah S. AlBadi & Mr. Talal Khalaf M. AlAghbari, who have exerted their efforts to make this book, including my other book "The Values of Well Being & its Secrets for a Better Living- Theories", with such a beautiful design and the way they were edited.

Everyone herein mentioned, has been providing me with fabulous encouragements that have helped me successfully writing my two books at the same time.

May God, the Almighty bless them and their Love ones with more Merits, Knowledge, Good Health, Happiness and Long Life.

Introduction

*Our Lord said "**We have indeed created men in the best of moulds**", **from the Quranic Version, ALTIN ((THE FIG)) - (95:4)***

Water is the elixir of our life. The purpose of drinking any liquid, water in particular, is to hydrate our body. There is a miracle in water as it is considered the most powerful medicine in the world.

Unlike the human made medicines (whether traditional or modern) it does not have any sign of side effects in our body.

The benefit that we will get from drinking Water will be great if we substitute the ordinary water with Hot Water. The various studies and efforts I have personally been exerting for the last three years have proven that by drinking Hot Water, it removes built up deposits in our nervous system. These deposits are responsible for creating negative thoughts and emotions. By removing these build ups, it will help to purify our thoughts and put us in a better emotional state.

Hot Water purifies the toxin, helps melting the fat deposits and destroys harmful bacteria in our body. It is the most important catalyst in losing weight and maintaining perfect body figure. It is a miracle liquid remedy that will keep us in shape physically and mentally. To be healthy, one must drink the required quantity of Hot Water in a day. Health is wealth, so we have to keep a healthy body to have a wealthy lifestyle.

Hot Water has been proving to cure various allergies, infection and diseases including those chronic ones.

However, remember who you are and how important you are. You are the most clever, expensive, value, genius, handsome (beautiful) person ***of yourself*** *in this whole World, simply because:*

"Human being is the most loved God's Creature among all His Creatures (Whether seen or unseen, known or unknown to Human himself).

His creation, composition and efforts, in contributing to the progress of human civilization remain one of the most marvelous God's miracles. The greatest among them is what has been revealed in His Books that He has descended from the heaven (the Torah, the Holy Bible and the Holy Quran). The books provide the duties, commandments, punishments and His blessings.

If we just know the contents therein, we will definitely discover the Miracle of God's love to His creature. Finding out His miracles to "Human" is simply discovering God's promised of the unaccountable blessings".

Faris R.S. AlHajri

Contents

Preface

And your Lord says: ***((Call on Me; I will answer your –Prayer–: but those who are too arrogant to serve Me will surely find themselves in Hell– in humiliation!)). From the Quranic Version, GHAFIR -((THE BELIEVER, THE FORGIVER - GOD)) - (40 :60)***

My beloved Mother (may her soul rest in peace) has lived all her life filled with health complications since I was two years old, as she remained suffering with the same disease for more than thirty years, she underwent ten surgical operations in her stomach throughout her sickness, among them, seven were done in Africa, and three in the United Kingdom. I am strongly confident that there are only few People in this World, who may have undergone such a large number of Surgical Operations in their whole life.

We are all from one Family, consisting of three Brothers and five Sisters, where I occupy the second to the youngest.

*Due to the strong suffering of my Mother from her chronic disease, we were all split in two different groups; the first one which consists of my eldest Brother (*Hilal*), and two of my Sisters (*Jamila& Alya*), were adopted by my Uncle (*Nasser*), while the other group consisting of two Boys,(*myself and Fuad*), and three of my Sisters (*Asma, Nassra& Amina*), were adopted by my Uncle (*Ali, *may his* *soul rest in peace), but our Brother (*Fuad*) was sometimes living with us, and sometimes with our Parents who were living in another Country. Therefore, the absence of our Parents has led to my thought that my Uncle (*Ali, *may his soul rest in peace) and my Auntie (*Shariffa, *may her soul rest in peace) were our Parents, so I was calling them (Dad & Mom). As my truly Parents came from travel from another Country, I was shocked to see them, the moment I was told they were my Parents, at that time, I was six years old, wondering* "how could I have two Fathers and two Mothers?!!!", *without being able to reach a contentment on the situation. Such visits by my real Parents were sporadic from time to time, but at a larger period of time due to the health condition of my Mother. However, we were not in position to see our Parents, unless after many years have passed, and at a very short period of time.*

My continuous prayers to God, the Almighty, to grant me the ability to cure my Mother from her chronic Disease that has lasted for a long period of time remained constant, but I could not achieve my dream. With the grace of God, the most Merciful and the most Grateful, she passed away at the time we were all near her during the last years of her life.

As my prayer continued, in order to find a cure for all Diseases and to all the People of the World, particularly, those Poor and the neediest, who do not possess the least financial ability to have access to medicine, treatment and healthcare, including those sick People, who possess chronic Diseases, and have lost all hopes to find a treatment for their Diseases, most especially, when I was suffering from some Diseases, such as Asthma, Rhinitis, low backache

(Lumbago) & Migraine, which remained for quite a long period of time, despite my frequent visits to the Doctors, the results remained negative, there were no hope for curing my Diseases. However, I intensified my efforts using all personal tests and researches in order to find a treatment for my Health complications, and with the Blessings from God, the Almighty, most Merciful and most Grateful, my wishes and prayers were finally accomplished in discovering the Miracle and Wonders of treatment from Hot Water.

God, the Almighty, has favored Human, as He made him the cleverest, the most genius and the greatest creature among all His Creatures.

He gave Human an intact brain in order to build Human civilizations, improving it, creating truelove, and understanding, so God has descended to Human religions and books from the Heaven; that is the Jewish, the Christianity, ending with "Islam" and His last book "The Holy Quran", all of that for the purpose of Religion, Knowledge, Wisdom and Culture to every Human, in order to access him to build strong Human civilizations throughout the History.

Due to that reason, the Arabian peninsula was in the first fate and the most in acquiring Human civilizations as it was the place where God, the Almighty, has descended all His religions from the Heavens, so it was there where the first Human civilization has started, and where as well Human started to wear cloth, made from animal skin, and then, Human civilization continued its spread all over the World.

But, due to the tough life environment Human has been

living in the Arabian peninsula, despite the tremendous efforts its leaders and scientists have exerted, but the absence of educational institutions in many parts, such as schools, universities and theatres, were resulting to negative impacts in spreading the education and culture. Therefore, the resources that the historic scientists and leaders were possessing were so much limited.

In the other side of the western part of the World, the environmental circumstances were very comfort in acquiring Human civilizations. So, schools, colleges, universities and theatres were built, resulting to easily spreading the education and the culture to those organizations, including the spreading of books, scientific and cultural events, moving from one generation to another.

Human civilization continued its erection in such a very fast manner, due to the increase number of scientists and intellectuals in various fields, mostly in the western countries. But destructive wars, fighting and conflicts between Human, were the results of the negative impacts to Human achievements in terms of scientific and technology development, hence, without the aforesaid, it could have been so tremendous much more than it is today.

With centuries passing, a majority of the Western People started to boycott the Churches, Families started to split in parts due to the absence of social life in most parts of the western society, while the eastern society remained grasping the religion essentials and the values of social life.

Due to the spreading of universities, colleges, schools, theatres and both, security and economic stability in the eastern countries from the twenty first century as well as the increasing of modernization, freedom, and the benefits from globalization, Human civilization has started to return back, as education and culture have strongly spread in such a very fast manner, in which most of the beneficiaries were those who exert their maximum efforts to develop skill, education and culture. However, copyright was strongly needed in order to protect Inventors and Writers, and with the blessings from God, the Almighty, and with the tremendous efforts by His Majesty, Sultan Qaboos Bin Said, in 2008, the Sultanate of Oman joined the World Intellectual Property Organization (WIPO), being a part of the World Trade Organization (WTO).

With the blessings from God, the Almighty, I got my share in acquiring a Copyright of my two books with the titles:-

- ***"The Miracle and Wonders of treatment from Hot Water".***
- ***"The Values of Well Being, and its Secrets for a Better Living"- Theories.***

Finally, I tender my sincere thanks and gratitude with a high pride to the Government of my beautiful country, the Sultanate of Oman, under the leadership of His Majesty, Sultan Qaboos Bin Said, on the benevolent life, stability, health, education, clean environment and all the favors that are being provided to both the Omani People and the Residents in the Sultanate of Oman, in which I have got my portion of such favors, that have resulted to my tremendous achievement I have made, and which will add another touch to the reputation of this lovely country in the world map on what the whole humanity in this whole world will benefit, as it is always a dream to every Human to accomplish a happy life filled with Health and wellness.

By Eng. Faris Rashid S. AlHajri
Technical Expert in the Office
Of H.E. the Undersecretary
of Ministry of housing.
Sultanate of Oman

A new Scientific Discovery achievement The Miracle and Wonders of treatment from Hot Water

Section I: My formula proving the Miracle and Wonders of treatment from Hot Water

The Main discovery behind this important event and my formula proving the achievement of this Scientific discovery, which will have a great impact to Human History:-

H_2O-means (2) Hydrogen and (1) Oxygen (ratio)= Water.

We need Oxygen to breath and to stay alive.

Therefore; water which contains both Hydrogen and Oxygen gives us live".

Water = Life

****** *The Sun produces heat to kill all harmful Bacteria, so that we can enjoy a healthy life and staying alive* **(Downes & Blunt).**

*** Section 0 – Principles of Health – Part 4b

Available at : http://www.pathlights.com/nr_encyclopedia/00prin4b.htm

(Accessed 21 May 2008).

Heat is the only natural element that destroys harmful bacteria, melts the fats and neutralizes the toxins that spread in our bodies from the food preservatives in our today's World.

#Sun=Heat(Hot)= Good Health

Therefore, Water+Heat = Life+Health

:i.e.Hot Water=Healthy Life=Miracle.

*** Oxygen burns Hydrogen in the living system, releasing energy that runs our bodies. Hydrogen is "the fuel of life".

*** Hydrogen & Oxygen

Available at:

http://www.tuberose.com/Hydrogen_and_Oxygen.html

{Accessed 6 June 2009}.

" If the Sun kills all Harmful Bacteria in the air, where as heat is the best source to neutralize toxins, destroy harmful bacteria and melt the fats, what about those accumulated ones inside our Bodies?!!!, and we all know that our Bodies contain toxins, harmful bacteria and fats which are definitely the main causes of the many and various diseases we acquire. However, the only mean to eliminate them, is through Heat. So, we really need to drink hot water in order to destroy toxins, harmful bacteria and fats completely inside our bodies. Therefore, this is considered a miracle".

Section II: My other formula proving the Miracle and Wonders of treatment from Hot Water

H_2O-means (2) Hydrogen and (1) Oxygen (ratio)= Water.

We need Oxygen to breath and to stay alive.

Therefore; water which contains both Hydrogen and Oxygen gives us life".

Water = Life

*** Hydrogen is the body's most needed nutrient.

Oxygen burns hydrogen in the living system, releasing the energy that runs our bodies. Hydrogen is "the fuel of life".

Water is formed when hydrogen is burned by oxygen.

Hydrogen is the ultimate antioxidant.

*** Hydrogen & Oxygen

Available at:

http://www.tuberose.com/Hydrogen_and_Oxygen.html

{Accessed 6 June 2009}.

Hydrogen= Good Health

Therefore, one portion of oxygen mixed with two portions hydrogen = water.

The only means to split hydrogen *with* oxygen *is by heating water which will thereafter burn the* oxygen *from water in a form of odorless and colorless steam, with no contents of any form of chemical elements.*

Hot water= (oxygen + hydrogen)+ Heat = Healthy life.

Hot water = Healthy life = Miracle.

However, by just drinking at least (8) eight glasses of hot water daily, with enough heat affordable by our body, we end up inhaling a large quantity of hydrogen, which is considered as the main body's nutrient element.

Section III: Benefits of hydrogen as stated in some Websites

Everyone knows that the body needs oxygen in order to live. So much emphasis has been placed on oxygen as the essential element allowing us to exist on the planet, that we tend to forget the other equally essential element namely hydrogen. Without hydrogen to combine with oxygen we wouldn't have water.

Oxygen burns hydrogen in the living system, releasing the energy that runs our bodies. Hydrogen is "the fuel of life".

Studies have shown that the human body stores hydrogen in its tissues. As we age, tissue-hydrogen depletion may lead to many of the symptoms of the aging process. This may cause sub-clinical dehydration since it appears that hydrogen may play a role in hydrating our cells.

Water is formed when hydrogen is burned by oxygen.

When we burn hydrogen in our cells, the energy that is released is used to run our bodies.

Hydrogen is the lightest and smallest element known to science. Due to its small size, hydrogen easily travels throughout the body.

No electron moves in the living system unless it is accompanied by hydrogen.

The hydrogen atom is the smallest of all the elements and has as much antioxidant power as the large, complex compounds.

Liver tissues store the most hydrogen, while the spleen stores the least. This is interesting in view of the fact that the liver is the body's first line of defense and needs a supply of the most antioxidants in order to do its work of detoxification.

Transport of hydrogen is the missing factor in the search of the cause of the aging process and the secret to aging reversal. As we grow older, our cells become dehydrated and our hydrogen pool becomes depleted. The hydrogen pool protects our cells from free-radical damage. Free- radicals are responsible for the aging process. There is a paradox in medicine, and that is the fact that oxygen is the source of all life and also is the major cause of aging. Much effort is being expended to find powerful antioxidants that may control or reverse cell damage by oxidative free- radicals. The single factor that is common to all antioxidants is that they are sources of hydrogen. Hydrogen is the ultimate antioxidant.

Hydrogen & Oxygen

The ultimate antioxidant

Available at:

http://www.tuberose.com/Hydrogen_and_Oxygen.html

{Accessed 6 June 2009}.

Albert Szent-Gyorgyi the Hungarian Nobel Prize-winning biochemist who discovered Vitamin C said that hydrogen is the fuel of life.

Our bodies store hydrogen in "hydrogen-pools" in the organs with the greatest amount stored in the liver which is the body's chemical factory and our most important organ for protection and self-defense. The liver detoxifies poisons to prevent them from getting into the body. Then the hydrogen is stored into the intestine, then the lungs, then the spleen.

Cancer changes the body's biological terrain in its favor by emitting toxins and using the body to dump its electrons in the urine and the blood becomes oxidized. Cancer cells have no hydrogen in them at all. Get lots of electrons into the body and you'll help reduce the cancer.

Hydrogen is the body's most needed nutrient. When you hydrate the cells they plump and become healthy and the body goes into an anabolic state – when the cells become dehydrated, the body goes into a catalytic state and eats its own muscles.

Eight glasses of good low surface tension water a day, exercise, good nutrition and natural diet truly are the basic keys to a long and healthy, happy life.

Hydrogen

What's hydrogen got to do with it?

What's the deal with anti-oxidants and free radicals?

Available at:

http://www.cybertown.com/slowaging3.html

{Accessed 6 June 2009}.

Hydrogen is the chemical element with atomic number 1. it is represented by the symbol H. at standard temperature and pressure, hydrogen is a colorless, odorless, nonmetallic, tasteless, highly flammable diatomic gas with the molecular formula H2.

Hydrogen is the most abundant chemical element.

Hydrogen/oxygen mixtures are explosive across a wide range of proportions.

In 1783, Antoine Lavoisier gave the element name hydrogen (from the Greek hydro meaning water and genes meaning creator) when he and Laplace reproduced Cavendish's finding that water is produced when hydrogen is burned.

The sun's energy comes from nuclear fusion of hydrogen.

Hydrogen – Wikipedia, the free encyclopedia

Available at:

http://en.wikipedia.org/wiki/Hydrogen

{Accessed 6 June 2009}.

Hydro: water, and genes: forming.

Named by Lavoisier, hydrogen is the most abundant of all the elements in the universe.

Hydrogen is estimated to make up more than 90% of all the atoms- three quarters of the mass of the universe! This element is found in the stars, and plays an important part in powering the universe through both the proton-proton reaction and carbon-nitrogen cycle. Stellar hydrogen fusion processes release massive amounts of energy by combining hydrogens to form Helium.

Hydrogen

(Updated 15 December 2003)

Available at:

http://periodic.lanl.gov/elements/1.htlm

{Accessed 6 June 2009}.

Section IV: Recommended methods for drinking hot water

The following procedures are highly recommended to experience the benefit of drinking Hot Water:-

1- ******One or two glasses of hot water, early in the morning, once you wake up and before brushing your teeth – <u>at standing position</u>.***

2- ***One or two glasses of hot water, after brushing your teeth, before having your breakfast.***

3- ***At least three glasses of hot water throughout the morning.***

4- ****** One glass of hot water at least <u>15-30 minutes</u> <u>before</u> <u>meal</u>.***

5- ***At least two glasses of hot water throughout the evening.
(Best if four glasses).***

6- ***One glass of hot water, before going to sleep.***

Notes:

*** (Should be given high consideration, due to its huge effects. For more details, please refer to section nine – paragraph (19), page "42").

- *The glass of water shall be approximate of 240-300 ml in volume.*
- *The Water shall be hot at a temperature of around (40°C) that means hot enough to feel it while drinking, but affordable without causing burn.*

In this manner, you end up being able to drink approximately ten glasses of hot water daily, where as this method would be of better effects than drinking the minimum required amount of eight glasses of hot water daily.

Never mind, just keep going to the toilet in order to discharge those harmful bacteria, toxins and fats, whenever is needed and much better once symptoms occur .

Never try to hold and refrain ourselves for a long time from going to the toilet, except in emergency cases. Remember, you are not alone who need to go to the toilet, there is a toilet everywhere! In the case of refraining ourselves to go to the toilet, this may cause constipation, which in return may cause the toxins to return back to the body's organs, become infected.

Section V: Results of drinking hot water

Drinking at least eight glasses of Hot Water everyday at a reasonable interval of time will be advantageous to our body. For us to acquire the wonders of hot water, the water should be at a temperature of around (50°C), where the hot will be felt by the throat and acceptable by your body. You will be lucky to discover the following miracles of hot water:

1. ***Prevent various diseases, symptoms and allergies which are the main causes of death of millions of people in the world. As a result, it will enable you to live a happy life by possessing a good health which you will consequently consider as a magical wealth.***

2. ***Heal people in pain, with sickness, allergies and diseases even if how critical it is. The healing applied to a new born and those of old age who are most of the time attack by Parkinson disease, skin deterioration, ageing, rheumatism, body pain and other health complication incidents of elderly period.***

3. ***Get rid of fat which are the main cause of***

being infected of many diseases, such as obesity, bronchial asthma, diabetes, hypertension, high cholesterol and other bad health condition due to excessive intake of fats.

4. *Improve brain memory that will develop you to be a genius. This wonder will lead you to total change of your personal, social and professional life to your self fulfillment.*

5. *Possess a good looking body and remove minor health complications such as, headache, body tension, colds, fever and back pain.*

6. *Acquire an exceptional personality or character due to stunning skin, attracted face and glowing smile which will influence you in dealing with people.*

7. *Have interest in peace, respect human rights and reject quarrels. Love others more than yourselves and enjoy daily life whatever you are doing or wherever you may be.*

8. *Realize the possible results of every incident before it will actually happen and to cope up complicated issues.*

9. *Love of natures and creatures for everything comes from our Almighty God.*

10. *Possess strong faith in God and believe that there will be countless blessings from God.*

Section VI: Results of drinking hot water in proper interval of time

** Week one:-*

- *Feel relax during bed time resulting to a comfortable sleep.*

- *Proper digestion which will discharge all our food intakes, up to (4) four times daily and continuous passing of urine. These changes will take for a couple a days until it will return back to the normal routine. The effect will differ from one person to another, between three months to six months or even more in rare cases.*

- *Don't worry, that means, your body is discharging those **Harmful Bacteria, toxins and fats**, which are the cause of diseases. Otherwise, the organ's mechanism will malfunction if all toxics will not be released since every organ has function that will keep you healthy, strong, energetic and inventive.*

** Week two:-*

The feelings of pleasurable sleep and relax body even if

sleeping was less than eight hours. Consequently, your body will ask you to compensate the sleep you lost, so do not hesitate or resist. Take a nap if possible, to feel the effect of drinking hot water and to satisfy "HIS MASTER", that's you.

The secret behind this reaction is that we are born in twins, "THE BODY" & "THE SOUL". Theory #(47) *as follow:*

"We have been created in two parts: The Body and the Soul.

Both parts are twin that if one contradicts with each other, there will be no harmony on their function. We have to be aware that the body and soul should agree on their purpose to human so in order to achieve a quality living on this beautiful world of paradise".

That is, your BODY needs his right to have a good rest, in order to give chance to your SOUL to perform his duties professionally & accurately as well.

Your Body, is happy, and continues to respond positively, keep going.

* *Week three:-*

Reduction of minor health complications, such as colds, cough and minor allergies. Improve the immune system, mostly in breathing and increasing your body energy, to avoid back pain and headache.

Your soul, is happy, and continues to respond positively, just keep going.

** Week four:-*

A recovery from chronic disease or any long term health complications, whether minor, or major.

A tremendous increase of your Body Sentiments & Feelings. You will start to become very sensitive, don't worry, your soul, is happy, and continues to respond positively, just keep on going. But your sensitiveness should be controlled in order not to hurt the feeling of others". Theory #(57) *as follow:-.*

"We should not hurt the feelings of others most especially those who look themselves as a weaker and emotionally depress individual. Hurting them will have a big impact on us for we will feel also the same pain that they are experiencing. Better to avoid hurting others so that we'll not end up in regretting on the wrong act that we did to others".

* The periods thereafter:-

You will start to acquire a tremendous knowledge and start to believe that "we live in a WORLD OF MIRACLES, surrounded by Miracles, they came from God the Almighty, the most merciful and most grateful, in order to protect us, and not harm or destroy us. We cannot see them, but we do strongly sense them".

You will start to love People, by considering every one as more value than yourself". Theory #(1) *as follow:-*

"We should never underestimate The Value of Human Being. Respect should always prevail. Any form of difference should be disregarded in dealing with human".

You will start to respect all "ALL TYPES OF CREATURES AND NATURE", you will start to love Peace. Theory #(43) *as follow:-*

"We need to cooperate with all human kind, whoever the person may be in terms of race, color, religion, size, age, shape, look, habits, power and wealth. Concern to nature, to all living and non-living things should be observed so that people will have a better place to live in, in this beautiful world of paradise".

You will start to fight for Peace, unity, cooperation, health, wealth ...etc, for all. Theory #(32) as follow:

"The more we are united, between all kind human beings despite whoever the person may be in terms of race, color, religion, size, age, shape, look, habits, power and wealth in all aspects of livings, the more we get stronger by the will of God and achieve a better living.

So let's benefit from every second of time that passed for definitely it will never come back, but our legacy will tell what we have remarkably done in the past".

You will start to hate War, Poverty, hatred, jealousy.

You will start fighting your Enemy through Peaceful means, using your Brain for protecting, preserving & improving peace, unity, cooperation, health, wealth ... etc, for all. Theory #(21) *as follow:-*

"We can never win a fight, battle or quarrel by using force, power and money, but we can win it by using our brain".

The best of all these, you will start to have a tremendous increase in FAITH (I repeat here Faith, and not Religious, hence all Human kinds are only united through Faith and not through Religion), You will be touched by your tremendous believing in the POWER OF GOD, OUR UNIVERSE AND THE CYCLE OF LIFE.

You will therefore, start to believe that NOTHING IN THIS UNIVERSE IS IMPOSSIBLE, EXCEPT BY THE ONLY AND LONELY WILL & POWER OF GOD. Theory #(25) *as follow:-*

"Nothing is impossible except by the will of God. We should focus on matters that have benefits to human, health, environment, peace and love. IT'S EASY; I repeat again, it's easy to achieve our goal and ambitions. The only secret in order for us to create, invent, discover, improve and solve is:-

- ***to live with love and peace in our heart***
- ***to learn how to handle tension, noise, anger, pressure and confusion***

In order to have a stress free and healthy lifestyle, try the following:

a) ***Wake up early in the morning after a pleasant sleep.***

b) ***Focus on greenly environment (trees, grass and flowers).***

c) ***Contemplate on shining blue color of the ocean and sky.***

d) ***Watch the sun when it rises and when it sets.***

e) ***Enjoy the golden sand at the beach and the glimmering stars on the sky.***

f) ***Appreciate the beauty of fog, snow, rain and hot weather on the desert.***

g) ***Listen on the flashing waves of the sea and droplets from the water fall.***

h) ***Love the people around us".***

*You will also start to feel the difference between FEAR & RESPECT . **Theory #(10)** as follow:-*

"We should remember that "Fear is to God and Respect is to Human". God is incomparable that we should be afraid of and human is His creation that we have the duty to respect".

You will start to identify the secret of our common dream. ***Theory #(31)*** *as follow:-*

"Every human has one dream:-

- ***To live and to fight for survival.***
- ***To be successful.***
- ***To be focused on the vision.***

The secret to achieve our own dream is to help others achieve their dream also".

You will start to find out the secret of forgetting our wrong pass. . Theory #(37) *as follow:-*

"We should not blame ourselves on any wrong decisions

we have taken in the past. No more regrets as long as we did it with good intention, believing that it is right with complete faith to God.

(((OH GOD YOU ARE SO GREATFUL AND MOST MERCIFUL WITH OUR PRAYER)))".

You will start to become Genius, by trying to innovate and Invent . Theory #(39) *as follow:-*

"When we are able to accomplish a very complex issue, we end up wondering how we did those things. We are surprised how we did it for sometimes things seem to be impossible due to the complex situations. And after sorting out issues we feel relieved and as if we have experienced miracles from above.

We can solve complex issues through the will of God. What we need to do is to: discover, search and invent the secrets in solving problems so that we'll have a wealthy, healthy and happy life.

And the secret to achieve the same is:-

a- ***Having a strong faith to God by seeking his blessings & mercy.***

b- ***Respecting human, nature, creatures and livings as per the cycle of life.***

c- ***Thinking always, focus on issues that are complex and solving it by:-***

First: Having a strong faith to God by seeking His Blessings & Mercy.

Second: Possessing a strong intention & tenderness.

Third: Planning by going through various experiments, thoughts, concentrations, studies, investigations, challenges, analysis, evaluations & discoveries up to our satisfaction.

Fourth: Execution by implementing what has been Planned.

Fifth: Repairing & Evaluation of the Defects occurred during the execution stage in order to reorganize them".

You will start to find out the secret of revealing your Shining Star, by just believe that, every one possess a Shining Star. Theory #(38) as follow:-

"Every Human being mentally fit, even if physically unfit, strongly possess a Shining Star that will definitely and easily turn him to be a great Genius in a specific issue.

And the only secret to reveal and discover the same is:-

Having a strong faith to God by requesting His blessing.

Eliminating and completely wiping out those negative mental attitudes which gradually destroy us by just fully being Optimistic.

Thinking always, focusing on issues that are found to be difficult to solve and not those easy one. Investigate, challenge, discover, study, analyze & organize.

Finally, the issue that is found to be strongly admired & attracted to both our Body & Soul will definitely turn to be a Shining Star that will completely change our whole life in order to achieve "THE VALUES OF WELL BEING & IT'S SECRETS FOR A BETTER LIVING".

You will be feeling that you are really reaching a successful life. But, remember to implement the factors of success in order to achieve your dream. Theory #(51) *as follow:-*

"In order to achieve a continuous success in everything we do, we have to consider the following factors of success:-

1- Strong Faith to God.

2- Intention and Willingness.

3- Planning.

4- Execution.

5- Evaluate the incidents and correction of the imperfections and defects".

Section VII: Personal researches done proving this Scientific discovery

1. *Pour **a glass of hot water** in a **plate** containing some left over **oily food**; the result is that water will definitely wash out the oil deposits. And then pour a glass of cold water in another plate containing the same as aforesaid, the oil deposits, will immediately turn into a sticky hard rubber look like component into the plate.*

2. *A toilet that has **no any sunlight penetration** and without proper ventilation, would definitely form very **harmful bacteria** that would result to a very bad smell. But, if you pour some hot water, without using any disinfectants, would result to the killing of some harmful bacteria, resulting to a better smell. Moreover, a toilet that has access to a direct sunlight would definitely smell much better, caused by the destruction of those harmful bacteria.*

3. *A **garbage** containing **left over food** for a longer period of time, will definitely cause spreading of harmful bacteria, resulting to a bad smell.*

4. *A plate containing fresh food left for a period of (24) hours in the kitchen, may not be spoiled. But, another plate containing the same food, if kept in the garbage tin, will definitely get spoiled in such a way it will produce bad smell; that is due to the accumulation of Harmful Bacteria in such a very easy manner compared to the first experimentation as aforesaid.*

5. *A left over* ***piece of meat stacked within the teeth*** *after meal, will definitely cause spreading of harmful bacteria, resulting to a bad smell.*

6. *We do always experience that, the Doctor do usually give the Patient stronger Medicines compared to those used previously, which really how smart those Harmful Bacteria are, hence, once the Patient is given a specific medication, the Harmful Bacteria will definitely split themselves in many various different parts, so that those ones who succeeded to remain alive, would turn to be much stronger than the previous ones in order to persist the medicines to the effect that, all the medicines that the Patient take will have no any effect to him after a specific period of time, if he could not be able to find the treatment of the disease, which would result to the suffering of the Patient waiting for his death.*

   Therefore, we shall remember how serious and dangerous Harmful Bacteria are considered to all Living Beings, including Human Being, as they tend to eliminate us, hence they are considered the most dangerous enemy of all Living Beings and how simply we may be able to eliminate them by simply drinking at least eight glasses of Hot Water as the best means to destroy and eliminate them before they destroy us completely. But, remember, we will never be able to terminate them from this whole Universe.

7. *We do* ***breathe a freshly pure smell air****, due to the effects of the sunlight that definitely, kills those harmful bacteria spreading in the air. Therefore, we may not live more than an hour if the sun disappears (****Downes & Blunt – 1877****).*

8. *The eyes sleep before the mind and wake up from the sleep before mind as well; which proves how important to drink at least two glasses of hot water immediately the time we just wake up early from our sleep; so that the brain system will perform it's duty by sending signals to all the parts of our bodies in order to perform their duties assigned to each one of them all over the day.*

9. *We all do experience* ***a bad mouth smell*** *at the time of waking up from a long sleep, due to the accumulation of harmful bacteria, toxins and fats coming out from the stomach, and the other organs in the body, before brushing our teeth, compared to the time we are awake even for a longer period of time where we do not experience such a bad mouth smell. And the reason behind this is that, Harmful Bacteria have strongly spread into the mouth due to its closure for a long period of time, which have resulted from the chemical reactions, our brain has generated in order*

to exert those harmful bacteria we have inhaled from the foods and the air we breath… etc. However, ***the moment we drink two glasses of hot water once we wake up and in a standing position****, the hot water exerts those toxins, fats and harmful bacteria, then completely remove them at the time we pass motion. But, if we do it in different way, by taking our breakfast immediately after waking up and brushing our teeth, the results become badly effective due to the accumulation of the toxins, fats and harmful bacteria which then remain inside our body, causing various health complications and diseases.*

10. *Some of the* ***evidences*** *proving how dangerous could be when we drink cold or iced water:-*

   a) *As we all know that our* ***constant body temperature stabilizes in normal conditions at around*** *(35°-36°) Centigrade, that means at a little warmer temperature, while cold water temperature of ranges at about (5°) Centigrade, resulting to the contradiction of the different "****electromagnetic current****" inside our bodies when we drink cold water, whatever higher the weather temperature would be, it is strongly advisable to drink Hot Water, and not the opposite, so that the temperature of the water we drink shall correspond with our body's temperature.*

   b) ***Cold Water causes the hardening of the fats inside our body's organs and blood arteries (vessels)*** *which thereafter, cause its coating to become thicker due to the increase thickness of the fats coating inside* *them including our stomach resulting to become big, as well as spreading all over our body's organs including our lungs resulting to breathing difficult leading to the attack of bronchial asthma. However, the complications become chronic within time which definitely lead to the acquiring of many health complications such as hypertension, cholesterol, attack of migraine, brain damage, stones in the kidneys and possibly causing its complete failure, and worst of all, acquiring cancer. Compared to the drinking* ***of hot water, which melts the fats, destroy harmful bacteria and neutralize toxins, which completely are being washed out in the toilet when passing motion****.*

c) *We all know that, the cause of* ***lightning & thunder*** *is when there is a big* ***difference in atmosphere temperature between cold and hot weather*** *at the time of* ***low weather pressure****. Therefore,* ***expect what may happen inside our bodies when we drink cold water where it's temperature contradicts with the one inside our bodies as previously aforesaid. Therefore, the results would definitely be dangerous, that may result to the failure of many of our bodies' organs.***

d) *We all, including every Creature in this world, do pass* ***motion and urine*** *that contain billions of harmful bacteria, toxins and fats causing to a bad smell. Whatever we may try to do, whether we bath water filled with the most expensive perfumes or pure milk. Whether we eat the most expensive* *food, fresh or even frozen food;* ***the smell would remain the same****.*

But, the ***miracle*** *behind all this is; a Person who* ***drink plenty of water****, his urine will turn to water look like and the motion to a softer loose like with a lighter color, both the motion and the urine will definitely have a* ***much better smell*** *compared to a Person who rarely drink water, or even substitute water with juices whether fresh or not, even if mixed juice or tea with water, the smell will definitely be very terrible, causing also constipation, due to the huge volume of harmful bacteria, fats and toxins within our body.*

We shall also remember that, the destruction of our health will gradually increase day by day, whatever treatment we do, and whatever medicines we take, whoever Doctor we visit, whatever sports and body exercises we practice. Remember also, the destruction of the same will be in a very faster manner, with those who tend to refrain themselves from going to toilets in a longer time compared to those who rapidly do the same by just starting to smoothly feel it.

Finally, I would also conclude the same as follow:-

"We live in a World filled with Miracles, and surrounded by Miracles. We cannot see them, but we do sense them, and they are here to protect us, not to harm us".

"Nothing is impossible in this World, except to return dead to life and except by the will of God".

"No Living being will ever live without water".

"No Living being, in time of famine, will ever survive by only food without water".

Section VIII: Diseases cured in People from Hot Water therapy between 2008-July 2009

1. Asthma
2. Hypertension (High Blood Pressure)
3. Diabetes Mellitus
4. Migraine & Headache
5. Anemia
6. Series of back pain
7. Urinary Calculus (Stones in the Kidneys)
8. Urinary Tract Infection
9. High Blood Cholesterol
10. Rheumatism & Arthritis
11. Stroke (Cerebra Vascular Accident)
12. Sexual and body weakness
13. Tiredness & Fatigue
14. Tonsilittis
15. Gastroenterisis (Food poisoning)
16. Insomnia (lack of sleep)
17. Colds, Flu & Fever

18. Heartburn
19. Ulcer
20. Constipation (difficulty in passing motion)
21. Parkinsonism (Involuntary Movement of the Body due to old age)
22. Hair loss (Baldness)
23. Skin Diseases
24. All Kinds of Infections
25. Alzheimer (defects of the Brain)
26. Heart Disease & Heart Abnormality since birth
27. Cancer (there is one case diagnosed and further follow up in other cases is being monitored).
28. Purifying and Regularizing Women's monthly Period.

Section IX: Proven Records from the Holy Quran

- *Being a Muslim, with a strong Faith to God, and having a strong believe and respect to all religions of God, to all Creatures, weather Human, Animals or Nature, as "**they were all created by God, the Almighty**" I went through various researches in the Websites, in Arabic language, to find out about "The Miracles of Water in the holy Quran", in which I was able to collect very interesting information on the same, therefore, I found out that, God has repeated about the Miracles of Water (63) times in the whole holy Quran, as follows:-*

1- *"Do not the Unbelievers see that the Haven and the Earth were joined together (as one Unit of Creation), before we clove them asunder? **We made from Water Every living thing, will they not then believe?***

- Sura Al-Anbiya "The Prophets" (21:30) -

2- *"Remember, He covered you with a sort of drowsiness, to give you calm as for Himself, and He caused rain to descend on you, from heaven, to clean you therewith, to remove from you the stain of Satan, to strengthen your hearts, and to plant your feet firmly therewith".*

- Sura Al-Anfal "Spoils of War, Booty" (8:11) -

3- *"Don't you see ye the Water which ye drink ?".*

- Sura Al-Waqia" The Event, the Inevitable (56:68) -

4- *"And we send down Water from the sky according (due) to measure, and we cause it to soak in the soil; and we certainly are able to darin it off (with ease).*

- Sura Al-Muminun "The Believers" (23:18) -

5- *"And we send the fecundating winds, then cause rain to descend from the sky, therewith providing you with Water (in abundance), though ye are not the guardians of it's stores."*

- Sura Al-Hajar "Stoneland, Rock City" (15:22) -

6- *"Seest thou not that God sends down rain from the sky, and leads it through springs in the Earth ? then He causes to grow, therewith, produce of various Colors; then in withers; thou wilt see it grow yellow; then He makes it dry up and crumble away. Truly, in this, is a message of remembrance to Men of understanding".*

- Sura Al-Zumar "The troops, Throngs" (39:21) -

7- *"And made therein Mountains standing firm, lofty (in stature); and provided for you Water sweet (and wholesome) ?*

- Sura Al-Mursalat "The Emissaries, Winds Sent Forth"(77:27) -

8- *"That sends down (from time to time) rain from the sky in due measure; and we raise to life therewith a land that is dead; even so will ye be raised (from the dead)".*

- Sura Al-Zukhruf "Ornaments of Gold, Luxury (43:1) -

9- *"It is He Who sendeth down rain from the skies,; with it We produce vegetables of all kinds; from some we produce green (crops), out of it We produce grain, heaped up (at harvest); out of the date-palm and its sheaths (or spathes) (come) clusters of dates hanging low and near; and (then there are) gardens of grapes, and olives, and pomegranates, each similar (in kind), yet different (in variety); when they begin to bear fruit, and the ripeness thereof. Behold ! in these things, there are signs for people who believe".*

- Sura Al-An'am "Cattle, Livestock" (6:99) -

10- *"And He it is Who sends the winds as heralds of glad tidings, going before His mercy, and We send down pure Water from the sky".*

- Sura Al-Furqan" The Criterion, The Standard" (25:48) -

11- *"Nor are the two bodies of flowing water alike,- the one palatable, sweet, and pleasant to drink, and the other, salt and bitter. Yet from each (kind of water) do ye eat flesh fresh and tender, and extract ornaments to wear; and thou seest the ships therein that plough the waves, that ye may seek (thus) of the bounty of God that ye may be grateful".*

- Sura Fater "The angels, Orignator" (35:12) -

However, when water is mixed with heat to become "Hot Water", its benefits become tremendous as hot water is the only element that destroys harmful bacteria, neutralizes the toxins and melts the fats inside our bodies. More over, it is the only natural element that has completely no side effects to Human, and would never harm him whatever the incident, as much as it is being consumed at an affordable temperature acceptable by our bodies.

Therefore, when Human benefits from the miracles of hot water, and the time he enjoys a healthy life, his body's activities become stronger, giving him a feeling of being active, and help him to increase his mental strength due to the huge amount of energy being emitted in drinking hot water .

Let me share with you my believe about what has been revealed by our Lord, the Almighty, In the Holy Quran, Sura AlWaqia "The Event, the Inevitable" (56), about the three main miraculous factors (FOOD, WATER & FIRE) that make us growing well, staying alive & enjoying an abundant health as follow:-

And drink Boiling Water on top of it (56:54)

Do ye then see?- The (human Seed) that ye throw out (56:58)

See ye the seed that ye sow in the ground? (56:63)

See ye the water which ye drink? (56:68)

See ye the Fire which ye kindle? (56:71)

Section X: Proven Records behind this discovery

The effects of heat and water; and its proofs on how being the best element that destroys harmful bacteria, neutralizes the toxins and melts the fats inside our bodies:-

1. *The experiments of cold, warm & hot water in the left over food plates containing some oil deposits "Fats" in the kitchen, in August, 2007.*
2. *The total cure of myself from asthma, migraine, lowbackage (lumbago) and allergy rhinitis diseases.*
3. *The bad smell of just a tiny piece of meat left in the mouth within the teeth, after meal, proving how harmful Bacteria, toxins and fats in our Body could be, being the main cause of the majority of the diseases we do acquire, since, there is no direct sunlight penetrating inside our body to destroy them.*
4. *Take an example of a closed room, mostly toilet, that has no any direct sunlight penetration, it would definitely result to a very bad smell, compare to a room or toilet that has a direct sunlight, it will always tend to smell much better.*

5. *The bad smell of any garbage from food, mostly, if left for a longer period of time.*

6. *Major Restaurants are provided with Hot plates, in order to prevent any communicable disease that may easily spread over, due to the sharing of same plates.*

7. *Remember, we always have temptation of catching colds in winter or whenever the weather is very cold, or even if exposed to a sudden change of our Body temperature from cold to hot, or vice versa, compared to summer, or when the weather is hot, we do rarely catch cold.*

8. *Infectious diseases may easily spread in tropical or cold countries if preventive measures are not properly taken, compared to those hot countries which do always experience a high raise of the air temperature, such as, the Gulf Countries.*

9. *Whenever we are exposed to a bad cold, the Doctor always advise to drink plenty of hot soups, warm water, and never eat ice cream, or cold food.*

10. *Whenever we pass motion, the smell is always terribly bad, even for those very rich people who do bath with milk, or use expensive perfumes mixed in their bath tubs; or even eating the most expensive fresh foods and fruits, that means, harmful Bacteria, toxins and fats in our Body are very dangerous.*

11. *If we do not brush our teeth for a longer period of time, or even we only brush once daily, and also just the time we wake up from our bed as well, the smell that is released from our mouth the time*

we breath, is of no doubt, very bad, that means, harmful Bacteria, toxins and fats in our Body are very harmful.

12. *Steam arising from hot water is being used to activate our blood cells, brighten our face, remove dead skin, clean and whiten our eyes, destroy harmful bacteria and neutralize toxins which are the main cause of skin damage, skin deforming that mostly appears to elderly people. However, steam is created when water is being heated, then the oxygen is burned, and the hydrogen remains in a form of steam.*

13. *The cause of thunderstorm and lightning due to the mixture of the air temperature between cold and heat in the atmosphere arising from the electromagnetic currents being emitted. That is, the heavy clouds that are in the upper atmosphere are extremely cold, then a low pressure in the air happens, the cold clouds are then pushed into the lower atmosphere where the air is extremely hot, the energy that is being emitted is huge, which means the hydrogen has produced a high level of energy.*

14. *If cold water is poured in a glass, then being emptied, and after that, if hot water is being poured in the same glass, high energy is being emitted, causing the glass to break in parts, due to the mixing of heat and cold.*

15. *Health experts warn to never try to take shower, swim or drink water except at warm temperature immediately after doing exercise, practicing any*

sport activity or even after coming out from sauna, due to the sudden mixture between the cold and the heat, that cause opposite electromagnetic currents, resulting to a possible heart attack, brain damage, or sudden increase in the blood pressure. Where as a permanent paralysis or even a sudden death may also occur.

16. *People are catching colds and flu, when they take hot shower, then suddenly going out where the weather temperature is very cold, due to the opposite electromagnetic current being emitted to the Human's body.*

17. *The main reason behind many People catching Rheumatism (pain of the bones), arise when they expose themselves to extreme cold being emitted from the air-conditioning, after taking shower, mostly the hot shower. However, it is strongly recommended to switch off the air-conditioning before having shower, including drying the skin completely with the towel and wear the clothes immediately after having shower, away from the air-conditioning.*

18. *The main causes of bad breath smell emitting from the mouth in the morning at the time of waking up, after having a long time of sleep, is due to the toxins, harmful bacteria and fats that are being concentrated in the stomach after being released from the body's organs, as the brain which is the main controller of all the Human's body organs, exerts a cleaning operation system of the body from the toxins which are really the main cause in*

creating diseases. So, this is really the main reason every Human suffering from a chronic disease, or even a person catching minor health complications such as colds, flu, fever, cough…etc. may not be able to enjoy his sleep due to the operations that are being done by the brain during the sleeping time in order to treat the Human's body from any disease or infection. Therefore, by just drinking at least two glasses of hot water in the morning while waking up and before brushing the teeth at standing position with a larger quantity at every time affordable by the body, this method helps the flushing of the hot water into the stomach in a faster way, the benefits resulting from the same, would be of great impact. The proof behind this is from the feeling of passing the motion immediately after doing this procedure, which means that harmful bacteria, toxins and fats are being washed away.

19. *A Person, who drinks water, finds himself acquiring strong appetite in eating. However, it is strongly recommended to drink a glass of water at least 15-30 minutes before meal, in order to create a balance in the food intake system, along with the many health benefits as stated by Dr. F.Batmangelidj.*

But here I would recommend substituting the glass of water with a glass of hot water. Hot enough as affordable by our body.

20. *A person, who drinks hot water, finds himself catching a tender sleep. Here, I would advise to take precaution in driving a vehicle or sitting in a quite place, by creating some kind of any*

movement that will make the body busy, otherwise the brain may end up getting confused thinking that it is sleeping time. Here the brain acts like a child as it gets confused in acting differently, due to the misunderstanding between "The Body & Soul". But, do not worry, as the brain is in a very comfortable living, since sleep is the king of our life, where as we cannot buy sleep at whatever price. Please also note that, Scientists and Health Experts, were not be able to cure the "lack of sleeping disease" despite various researches and tremendous efforts that have been carried and still being carried. So, you are really lucky, that you were able to cure yourself without any cost or physical implications. More over, without endangering yourself in taking sleeping peels or any other form of toxins, except yourself in abiding to drink hot water frequently, that has no any side effects, whatever the quantity you drink.

Section XI: Foods and drinks that are advised to take and those to eliminate

First, foods and drinks that are strongly recommended to eliminate in order to benefit on the miracles of hot water:-

- *All types of soft drinks, as they are all made from toxic materials, and they do not contain any natural materials.*
- *Alcoholic drinks, whatever possible, which are considered having containing strong toxins that are very harmful to Human.*
- *Packed juices, unless in some urgent needs.*
- *Huge quantities of meats, unless in some occasions such as parties, but in reasonable quantities, and it is strongly recommended to eat meat not more that twice weekly.*
- *Canned foods, unless in some urgent times.*
- *Heavy intake of foods at a time.*
- *Fried foods, unless in some urgent times.*
- *Ice cream in frequent times, unless in urgent needs.*

- *Very cold drinks.*
- *Cold water.*

Secondly, the foods and drinks that are strongly recommended to intake in order to benefit on the miracles of hot water:-

- *All types of natural drinks such as fruit juices, whenever possible.*
- *Fish at least three times weekly.*
- *Fruits and vegetables in a daily portion, at least two types every time.*
- *Fresh foods and organic foods (foods that are being packed, but made from organic material, without any form of preservatives, additives or toxic materials).*
- *Milk and dairy products including different types of nuts.*
- *Hot water*

Section XII: Some of the researches made by Dr. F. Batmanghelidj & Downes & Blunt

"You're not sick; you're thirsty. Don't treat thirst with medication."

Our life, our planet. Over 70% of the earth's surface is water. Water is the basis of all life which includes our Body.

You are not just what you eat; you are what you drink.

This is why water is so important to your health.

Water is the basis of all life and that includes your body. Your muscles *that moves your body are* 75% water*; your* blood *that transports nutrients is* 82% water*; your* lungs *that provide you with oxygen are* 90% water*; your* brain *that is the control center of your body is* 76% water*; even your* bones *are* 25% water.

Our health is truly dependent on the quality and quantity of the water we drink.

All animals drink plain water. If you watch a cat or dog, it will always drink water and wait approximately half an hour before eating food. The exception is when they receive a treat. They take it and eat it then or eat it later.

This sequence is critical to maintain good health. When you drink plain water and wait a half hour before eating food, the water goes into the stomach and triggers glands to release a chemical (hormone) to coat the lining of the stomach to protect it from the hydrochloric acid that will be released when food is put into the mouth. The water and chemical (hormone) then go into the small intestine where the chemical stays, waiting for the food, which will be coming down later. The free water leaves the small intestine and goes into all the cells of the body to fill them up (hydrate), top them off like watering a garden. After the cells are watered, the remaining water is pumped back into the stomach to water (hydrolyze) the food. Our energy does not come from the food we eat, but comes from the water that puts the hydrogen (atom) in the food. We run on hydrogen. The food we eat gives the vitamins and minerals we need to stay healthy. If you don't drink plain water (1 or 2 glasses) before eating food (waiting 15 to 30 minutes), then the body must borrow the water from itself. If it borrows water from the blood, then the arteries must draw up (constrict) and the heart must pump harder to pump the thicker blood. We call it high blood pressure. If the body borrows the water from the brain, you get a dull headache or a headache that can turn into a migraine. If you drink alcohol and it pulls water out of the brain, you get a hangover headache. People who live in cold countries put alcohol in their gas tank to pull the water out of the gas so it won't freeze.

Wherever the body borrows the water from, it compromises that organ or tissue, causing pain or an abnormal function like constipation. Constipation is when the body pulls the water out of the small intestine and colon, causing the body waste to stick to the walls of the intestine and colon. When your stool is moist, it slides out with no effort.

When we breathe out (exhale), we lose water. If you blow on a mirror or glass window, you see the moisture (fog). When the body borrows water from the lungs, you may get shortness of breath, leading to asthma and other breathing problems.

The lungs and heart are very close to each other so if the body pulls some water from the heart (muscle), you may get angina pain that may, down the road, lead to a heart attack.

Muscles need blood and water during exercise, so if you were to eat food and walk or climb stairs when low on water (dehydrated), you might get shortness of breath, have an asthma attack, get angina pain or leg cramps. They say never go swimming after eating because you may get muscle cramps.

When in the digestive mode, the body diverts blood from the arms and legs to the intestines to process the food.

Certain foods are easy to digest and are loaded with water, such as fruits and vegetables.

Sometimes a person might have a pain in their back, warm or leg just go away because of the water and salt. It gets into the blood in just a few minutes and will neutralize the acidic tissue when tissue becomes too acidic from lack of water and salt. It gives us a pain signal telling us to drink some water and take a little salt.

The greatest medical discovery ever made was by F. Batmanghelidj, M.D. (They call him Dr. Batman.) He discovered that the human body runs on hydrogen from water and if we don't drink enough water before we eat food then the body must borrow the water from itself. And wherever it borrows the water from, that is where you will have medical problems down the road.

Dr. F. Batmanghelidj

WaterCure – The Miracles of Water to Cure Diseases

Available at :

http://www.watercure.com/

(Accessed 17 May 2008).

*** The sunlight is a most marvelous system, and without it you could not remain alive an hour. It enables you to have strong bones, teeth and nails.

Every living thing in our world is dependant upon the sun. without sunshine, nothing could live.

In 1877, two researchers, Downes and Blunt, discovered that sunlight can destroy harmful bacteria. Today, it is used to treat bacteria infections. Sunlight on the body dramatically lowers high blood pressure, decreases blood cholesterol, lower excessively high blood sugars, and increases white blood cells.

(Downes & Blunt - 1877)

*** Section 0 – Principles of Health – Part 4b

Available at : http://www.pathlights.com/nr_encyclopedia/00prin4b.htm

(Accessed 21 May 2008).

Section XIII: The History behind this Scientific Discovery

My Mother has suffered from a chronic disease for more than thirty five years since her childhood until her death in 1993. She had undergone ten operations among them seven in Africa, and three in the United Kingdom, resulting to a brain evolution of my soul, that I should find a cure of all diseases in sick people in this whole World and disappearance of poverty without any discrepancies or racism between people. Thereafter, I started acquiring a strong believe that, God the Almighty & the Greatest of all, did neither create diseases, nor poverty. But, it's we Human being, who have caused the birth of so many diseases and poverty, due to our selfishness.

From 1988 : Just one year after joining my work, after my high level diploma graduation from "Oman Technical College", I started acquiring a series of serious colds, prolonging to a longer period of times, up to three months, no medication was found viable, despite my several visits to hospitals & clinics.

The symptoms persisted for a long period of time. I was also very workaholic, working daily for a minimum time of fifteen hours, up to a maximum time of twenty three hours, without a break, without food intake, without water intake, no time to loose, as I was always rushing to finish my works, and would never leave my office, unless I finished my work.

1993 : I was highly interested in experimenting everything, searching for the unknown, including using my body in various experimentations in order to find out a permanent cure of my cold allergy, such as, but not limited to, the using of toothpaste, disinfectants, salt, different herbal leaves inside my nose.

in 1998 : My work was involving preparation of tender documents as well, and the only technology I was allowed to use, was a copier machine, connected directly to my PC, through cable only, no network system was available at that time.

There was no more staff in my section, as all of them, were foreigners, Quantity Surveyors, terminated after the country economic crisis of 1985 due to the collapse of oil world price, while the only Omani colleague, a lady, civil engineer, graduated from UK, had to take the advantage of the early retirement, temporarily introduced by the Government .

Suddenly, from 1997, a lot of tenders, were requested to be issued, through tender, in which I was requested to prepare and issue tender documents, for construction of Government housing units, for low income families.

One day, I was praying in my office, it was evening time, as my office was very squeezed, the only available space, was just in front of the copier machine, there was a continuous copying of thousands of prints, I felt like chili being pouring in my nose, resulting to a very terrible cold, which has lasted to more than three months, without finding the cause, or the treatment.

One day, I read a health news article from the newspaper, mentioning about the danger of ammonia chemical, mostly found in copier machines & printers.

The only solution was to separate the copier machine from my office deck, through a partition.

In 2003 : A sudden acquiring of asthma disease, grade 2 level, was discovered in me in Royal Oman Police Hospital, Oman, by Dr. Mathew, not curable, but can be controlled only if proper medication is taken. Prescription of inhaler (a spray used by asthmatic People) was advised, but the same was being addicted by my body, that is lesser time interval was experienced.

One month after, I was advised by my wife Gloria, a BSc. In nursing with an experience of the same for more than twenty years, to discontinue the medication, an alternative one was provided to me from "Muscat Pharmacy", a quite expensive one, consisting of a capsule containing powder, it was much comfortable, and the interval was around (4) days.

But, later on, again my body started to get addicted to it, resulting to it's demanding at lesser interval time, no choice, I had to continue the prescription, and started to cope with the disease, by calling it "My Friend and my partner, so please don't disturb me, I have accepted you as my partner for life", so that I could be able to enjoy my daily life.

In 2004 : My wife was very interested in doing more

researches from the Internet, in which she found an alternative medication, that was "Anti-Histamine, Promethazine, for anti-allergy", made from syrup in a bottle, it was looking and tasting like the cough medication, cheap & very comfortable, I had to take it nightly, before bed time. But, at the beginning, I had to use the inhaler capsule, whenever I caught cold, or exposed to dust, especially old dust.

I then, became very interested, along with my wife, in finding alternative medication, I was continuously receiving health news via e-mails from various websites, reading daily newspapers, mostly regarding health tips in which there was an article about "Water Therapy", mentioning about drinking at least, 1.5 Liter of water at room temperature, early in the morning, best before brushing teeth.

And, in another health news, from the newspaper, there was a news about the danger of drinking "cold water", mentioning that it causes concentration of fats inside the blood vessels, which would result to blood blockage and definitely, causing hypertension, up to the extent of causing heart attack, assuring previous news from "The Times of Oman Thursday magazine" I already read since 1995, in which I stopped drinking cold water including my family as well as banning of cold water from the refrigerator.

A full cooperation from my wife was provided, resulting to everybody, including my two kids, in drinking warm water, so whenever any guest at Home was requesting to provide him cold drinking water, we were just getting the warm water and adding ice on it, in order not to disappoint our Guest.

Due to my weak Immune System, and continuous working for long period of times, as aforesaid, I collapsed, and totally lost control of my body, luckily, my Wife was there at home, she didn't panic, despite myself insisting to be rushed to the hospital, she then carefully helped me walk to the bed, where she laid me down facing upward, with my head leaning down to a lower level than my body. Within few minutes, Dr. Ayleen, our neighbor and friend, and a Consultant from the Royal Hospital, Oman, arrived including my younger brother, Fuad.

Within about twenty minutes, I was back to normal, and my wife kept warning me about the danger of workaholic, I didn't take it so serious, later after a month, the attack was repeated, but that time it was in my office, where everybody had already left.

Due to my knowledge of the first aid tip, I just slowly walked to the Sofa, and did the same practice; I was then back to normal condition.

In July,2007: I was badly infected with a serious infection of the throat, while ending my vacation in California, I had to be rushed to the nearest health centre, Badr Al-Samaa Polyclinic, once I arrived along with my family in Oman.

In August, 2007: Randomly, while washing the dishes, after lunch, I realized that, the fats from left over food, were sticking in the plates, then I used hot water, in which I discovered the plates were completely shining, by only using hot water, without soap, it was amazing news to me, in which I realized that the only mean to destroy harmful bacteria, neutralize toxins and melting the fats from our body, was to substitute the warm water to hot water.

I also wondered, why just a small tiny piece of meat, stuck in the teeth, was badly smelling, while, if a whole plate of cooked meat left at room temperature, for a longer period of time, say four hours, can still be consumed.

Therefore, our body consists of very harmful bacteria, toxins and fats, resulted from the motion inside our stomach, that is so dangerous, the longer they stay inside our stomach, and the only and best means to wash it out is through the consumption of fresh hot water, continuously.

I therefore, immediately, started the experimentation to myself, which, definitely resulted to a continuous passing of motion and urine as well, and I had to stick to my decision, despite some confusing research news, among them, contradicting previous ones about water, worst, when one of the researches, denouncing in such a warning means, about the danger of drinking plenty of water, which I completely deny such news.

I started feeling a tremendous improve of my health condition, mostly with my immune system, as it was getting stronger and stronger.

In January 2008: While visiting the doctor, for a small cold symptom, I was denied by the doctor, that I had asthma, and that my breath was in a perfect condition and I was looking very healthy, mentioning that it was just some little colds.

Again, in another visit, I was informed the same news.

No more "wizyness" was even experienced, I personally was tremendously improving in all means as aforementioned.

I then, recently, turned into a different Person, in the following manners:-

1. *Highly interested in creating philosophical theories in proportion with the events being happening in my daily life, in which I successfully achieved to register them for copy right under the title:-* ***"THE VALUES OF WELL BEING AND IT'S SECRETS, FOR A BETTER LIVING", where as a separate book was being prepared for its issuance along with this book .***

2. ***Acquiring*** *a very sensitive feeling, that has led to the following issues:-*

   a- Highly interested in making peace.

   b- Highly admiring a peace & a quite environment.

   c- Do not like to watch news, movies or see events connected to any type to abuse, killing, hatred, quarrel whether to human or to any living creature, but I only remained in believing in the cycle of life, such as slaughtering of beef for the purpose of eating the meat, not for any other purpose, such as causing any form of suffering of any animals for the purpose of fun, sport of competition.

   d- Becoming very emotionally affected if quarrelling with someone, mostly to those who see themselves as weaker than us.

3. *Having a strong believe that,* ***"NOTHING IS IMPOSSIBLE, EXCEPT BY THE WILL OF GOD, SO EVERYTHING IS POSSIBLE, EXCEPT RETURNING OF LIFE AFTER DEATH", "WE ARE LIVING IN A WORLD OF***

MIRACLES, SURROUNDED BY MIRACLES, THEY ARE THERE FROM GOD TO PROTECT US AND NOT TO DESTROY US".

4. *Highly interested in issues that are considered to be very complex, in all manners of life, weather at work, at home or elsewhere, in order to find out it's causes, it's sources and the way to solve them, and finally, to find out PREVENTIVE MEASURES on how to prevent it to happen again, or how to deal with it in the future .*

5. *Very interested in working with very intellectual people and with all type of Human being as well, in order to acquire more knowledge, by strongly believing on "THE SECRET OF ACQUIRING MORE KNOWLEDGE.* ***Theory #(42)*** *which states as follow:-*

   "The best way to acquire more knowledge is through sharing our little ones with others openly".

6. *Very interested in reading books related to self being and improvement.*

Section XIV: Some advises I would like to provide to the readers of this book

A hope from hopeless has strongly returned in Human's life after centuries of dissatisfaction.

Human has been in race of wars fighting his dangerous enemy, "diseases", discovering the traditional medicine from the Stone Age. But, side effects from the herbal medicines were more than its effectiveness due to the strong chemical they contain not in proportion with the Human's body.

As the race of wars continued, Human discovered a new hope of health living, thought to have defeated his dangerous enemy, the moment he invented the modern medicine.

Such hope has been gradually degraded due to its huge side effects from the high volumes of toxins they contain, until the modern life of today.

As the race of wars got stronger in today's world, and the Human dangerous enemy started to attack our Youths, our new generation has been involved in the race of wars.

The world got divided into two parts.

One part strongly recognizing to adopt the modern world, including insisting to stick with the modern medicine, completely denying the ancient world, including their complete denial of the traditional medicine.

The other part strongly recognizing to adopt the ancient world, including insisting to stick with the traditional medicine, completely denying the modern world, including their complete denial of the modern medicine.

My message to the World is: "THE WORLD IS CHANGING FOR GOOD".

The race of wars between Human and his dangerous enemy "Diseases" has now come to the end by discovering "the Miracle and Wonders of Treatment from Hot Water".

Remember, our bodies reject strangers and only traditional and modern medicine shall be adopted in emergencies cases when the curing from hot water drinking has failed where there is no improvement to the patient's health condition. But, remember always "PREVENTION IS BETTER THAN CURE".

Remember as well, the number one enemy of Human is human himself.

The most dangerous enemy of Human is disease.

The most loved creature to God, the most merciful and powerful of all, is Human.

The most clever creature God has ever made, is Human.

The number one enemy of God is devil, as he takes advantage of those who lose faith and their contacts with God, in order to destroy the most clever, loved, genius creature preferred by God, who is "Human", by just creating compulsion and wars between Human to Human.

Therefore, all Human kinds shall strongly return their faith to God, by calling upon His name, everywhere, every time.

As God is the one who listens and knows everything in order to defeat the devil.

As God said: "And (if at any time) an incitement to discord is made to thee by the evil one. Seek refuge in God. He is the one who hears and knows all things"-from the Quranic version FUSSILAT "EXPLAINED IN DETAIL"-(41:36).

No Human can ever predict the future, only his emotions are carried away by his thoughts.

The only secret is that; only when Human being gets stronger faith to God, his emotions are being carried away by his thoughts in closer to reality.

But, when he loses faith and turns himself into immoral, then definitely his emotions are being carried away by his thoughts, closer to misleading.

I finally advise every one who doesn't believe what

has been revealed in this book about the "Miracle and wonders of curing from hot water", and its huge benefits to Human himself who abide to drink hot water, whether sooner or later, that there may be a time where he may get sick (here I do not intend to warn, but it remains a normal incident, as there is no Human in this whole World who doesn't get sick, even a temporary sickness, such as infection or else). And at the moment he just drinks medicines (whether Modern or Traditional) to reduce the pain he may have been gone through, the medicine may not do its functions, due to the strong chemical they content, which are considered as dangerous toxins that Human has made, that is "the medicine", even thought, they may be successful in curing many diseases, infections and allergies that attack Human, but, they remain harmful to Human himself in longer period of time.

And a the time he drinks at least one or two glasses of hot water, he ends up feeling that he has inhaled an "incomparable medicine" that has been created by God, that is the water itself, that is colorless, odorless, no harmful, as it is the most marvelous in our life system, and without it no living being can survive.

Water is there everywhere, whether from the rain, canals, rivers, water falls, lakes, water between rocks from mountains or even wells, as much as it is potable water. In the occurrence where there is no potable water, or in emergencies cases, such as wars, natural calamities and famines, I would strongly recommend boiling it, or exposing it to the sun, in order to kill any harmful bacteria that may be existing it becomes purified and can be potable.

I strongly recommend all the mothers to give their babies hot water with the same temperature of the mother's milk or any other type of milk, and this is the prove, as mothers never give cold milk or even the milk at room temperature to their babies, otherwise, it causes stomach pain to their babies and some other health complications. So, the same principle shall be applied when giving water to their babies, it should definitely be hot with the same temperature of the milk, isn't amazing? Definitely, IT IS !!!

Whenever you experience any symptoms of any health complications, whether allergy, or sickness, I would strongly recommend you to increase your intake of hot water. Keep away from the medicine as much as possible, unless in emergency cases, or in urgent requirement. Keep also your intakes of the steam arising from hot water. You can use the simplest method in doing this by just fill a third of the electric water boiler, cover your whole body with a blanket, then heat the water to a boiling point. Keep the cover of the electric boiler opened. But, keep on controlling the water temperature, by simply switching off and on. Make sure you do not cause any burn to yourself. Use this method for about ten to twenty minutes. Inhale the steam from your mouth as much as you can, then release it through the nose, and vice versa. Then, inhale from one side of your nose, by closing your mouth, and then release it from the other side of your nose, and vice versa.

Remember, in the case you do not possess an electric water boiler; you can also use another simple method. By simply boil the water in a proper container, heat some stones with fire, then pour one by one into the pan while covering yourself with a blanket. Inhale (breathe in) and exhale (breathe out) the steam in the same process of the electric water boiler. Moreover you can also benefit the use of sauna room whenever available with the same breathing process of the electric water boiler.

I would always recommend doing this method, after shower. Never take shower immediately after the steam method, just dry yourself with a dry towel. And if necessary, only take shower after at least half an hour of having your steam.

By doing this method, you are helping activating your blood cells, becoming more energetic, brightening and softening your skin and feeling very active. Simply, because of the high quantity of hydrogen you have inhaled, which is considered the chemical element with atomic number 1, and being represented by the symbol H, helps increase your "hydrogen-pools" that exist in your body organs with the greatest amount stored in the liver which is the body's chemical factory and our most important organ for protection and self-defense. The liver detoxifies poisons to prevent them from getting into the body.

When the cells become dehydrated, the body goes into a catalytic state and eats its own muscles.

Some people misinterpret the miracle and wonders of treatment from hot water in the same manner of the treatment using man made medicines (whether traditional or modern). By the moment they become cured from any health complications through the method of drinking hot water, they discontinue this method. This is definitely wrong!!!

I strongly advise every one who really intend to live a healthy life, never stop drink hot water as much as he is still alive.

Unless in emergency cases, or out of control. But here, you are the loser. Whenever there is no possible access to hot water, such as during wars or natural calamities, I would strongly advise to expose the water in the sun, even if contaminated. With this method, the ultraviolet invisible rays from the sun will be well enough to purify the water, where as the harmful bacteria will get destroyed. The water from the sun exposure shall be consumed by the moment it is removed.

I have previously revealed in this book that, a person drinking hot water, will lead him acquiring a tremendous talent. The evidence is simply by finding out a person becoming completely asymptomatic, with no any health complications after his complete benefit from the miracle and wonders of treatment from hot water, he ends up acquiring a tremendous amount of energy inside his body. In this manner, the brain being as the "SOUL" becomes completely in full relaxation away from its worries of trying to cure the "BODY" from any disease. However, instead the brain to remain away from thoughts, the person who is the beneficiary, starts thinking about the future, getting closer to ethical issues to increase his faithfulness. He then gets involved in contributing to peace making, building the human civilization, completely denying any issue that intend to harm human kind, trying to help raising the living standard of others as a sign of his defiance to poverty, having the intention to solve public issues as a denial to wars and detestation, loving of all types of nature, respecting all types of creatures whether known or unknown, seen or unseen to human himself, except the Devil.

Every human being search for a good quality health and exerts all his most available efforts in order to achieve the same. Whatever his age, whether he is baby, child, adolescent, adult or even old aged. However, hot water shall be consumed by all human kinds, despite of his age, color, religion, doctrine, size, shape, look, and appearance, financial or even physical potency. Just by being a human being created by God. Not the one made by human, such as the human robot.

People suffering from obesity are due to the high amount of fat deposits in their bodies. Obesity is considered the biggest threat to human health as it is the main cause of hundreds of millions of people around the globe suffering with various chronic diseases, such as diabetes, asthma, blood hypertension, ... etc. People, who abide the method of drinking hot water, benefit in losing their weights in just a short period of time in the same manner being described in this book. Most especially, those who practice body exercises at a rate of just ten minutes daily. Even those who for whatever reason could not be able to practice body exercises, they still benefit to lose weights by simply abiding the method of drinking hot water. But, the results to people who practice body exercises would definitely be much better than others who do not so. For that reason, I would strongly recommend you to practice the same with even those light body exercises such as walking of at least ten minutes daily. Just do it, for you to see the astonishing results.

Drinking hot water helps the cleaning and purification of the kidneys in order to get rid of any impurities such as kidney stones. It helps maintaining women's monthly menstruation as well, simply as it purifies the blood, the blood cells, the kidneys and the overall body. Remember, by drinking hot water, the blood turn to pure red color look like, unlike people who do not drink hot water, their bloods have a tendency of having the color of dark red close to brown look like.

People who drink plenty of hot water of about eight to ten glasses per day were experiencing an increase of their body weights, arising from the plenty of appetite that occur from the same. Simply because they do not drink a glass of hot water at least 15-30 minutes before having their meals. I strongly recommend to have a glass of hot water before meal as aforesaid, in order to keep a control of the body's weight from one part, and to benefit from this method as previously mentioned in this book, most especially where Dr. F. Batmangelidj has strongly recommended to drink a glass of water at least 15-30 minutes before meal, in order to create a balance in the food intake system, along with the many health benefits. So, here I just recommend substituting the glass of water with hot water.

Some People have misinterpreted about the method of drinking hot water as they tent to temporarily suspend this method whenever the weather gets hot or in summer periods, or they get exposed to the heat of the sun, or if they work in hot environment, or even if they practice sports and body exercises.

Remember, these important tips as follow:-

1- ***Our body's temperature never go below 35 degree Celsius, nor even exceed 37 degree Celsius under body healthy condition, from the moment we get life inside our mothers' wombs until our last breath where death take us away from this world.***

2- ***The first nutrition we have ever intake from the moment we were inside our mothers' wombs to the first milk we have taken from our mothers after birth, including the first bottled milk we have ever took, all these were hot above room temperature, that is above 25 degree Celsius, most probably at a temperature of around 35-37 degree Celsius, same as the human normal body temperature.***

3- ***I would just like to remind you what has been revealed in this book about the side effects of drinking cold water, including other types of very cold drinks.***

Nevertheless, anything that is cold, or even contradicts with our body's temperature, will have a tendency to destroy us by causing failures to our body organs.

So, no matter the weather is, whatever the temperature of your surrounding environment would be, I strongly recommend you to abide the miracle & wonders of treatment from hot water.

Some other People as well have misinterpreted about the method of drinking hot water as they tent to add some other elements into a glass of hot water, such as pure honey, or squeezed fresh lemon, or salt, or green tea, … etc, because they don't like the taste of water, or pertaining to cure various allergies and diseases. That may not be wrong at all. But, here by doing so, this method is definitely considered as "HERBAL TREATMENT" or in other meaning, "TRADITIONAL MEDICINE".

Everybody knows that all types of man-made medicines, whether modern or traditional, help in some way, but may sometimes cause side effects in another way if wrongly consumed. Both, in short and long term period of time. Simply because our bodies are very sensitive. Nevertheless, THERE IS NO SUBSTITUTE TO WATER. As the molecular structure of water is (H_2O), anything that is added to it is no longer water and immediately changes its credibility. IT WILL SURELY BECOME NOT WATER ANYMORE. BY THE MOMENT WATER IS MIXED WITH ANY OTHER ELEMENT, OR EVER THE SAME ELEMENT IS ADDED EXTRA TO WATER, IT IS NO LONGER WATER AT ALL. Do you believe this? I strongly do. THIS IS TRULY A MIRACLE.

Let me share you some of the following evidences as follow:-

A cup of black tea is composed with around (2%-5%) pure tea and (95%-98%) water, that is what I have assumed, as the real ratio may vary. But the miracle is that our body will never accept to substitute pure drinking water with black tea. Otherwise, we end up having urinary infection, stones in the kidneys, and much other health complications with time passing.

Furthermore, with other elements, such as the freshly squeezed lemon, pure honey... etc. By mixing any of such elements, it definitely becomes as juice, it is no longer water. We cannot at any mean substitute fresh water with "juice"; otherwise we will definitely experience some health complications.

Finally, let me share you the following evidence proving that "Water" which is known to the modern science as a mix of two hydrogen atoms and one oxygen atom with the chemical symbol (H_2O, or hydrogen monoxide); is a MIRACLE element:-

- ***If we add another unit of oxygen to water it becomes "Hydrogen Peroxide" with a chemical symbol "H_2O_2". Here, it is no longer water, it becomes another element. If consumed, it will definitely cause harm to our health. As it is only made for the purpose of disinfecting our kitchen and bathroom, and not intended to prevent, treat, or cure disease conditions or to affect the structure or function of the body.***

  ****** H_2O_2 is simply the scientific name for Hydrogen Peroxide. It is naturally occurring water-like liquid that has many practical applications both inside and outside the home. Hydrogen Peroxide (H_2O_2)is made up of two hydrogen atoms and two oxygen atoms. H_2O_2 looks like water (H_2O), but that extra oxygen***

molecule makes this natural water additive one of the most powerful oxidizers known to man. It is formed in nature by the action of sunlight on water, and even in honey.

You are probably already familiar with using the low grade 3% hydrogen peroxide. Being a powerful oxidizer, hydrogen peroxide kills bacteria, viruses and fungi on surfaces. This means it is also great for disinfecting your kitchen and bathroom. You can make your household cleaner and safer just by substituting hydrogen peroxide for those caustic chemicals you are currently using.

Technical Grade Hydrogen Peroxide is not intended to prevent, treat, or cure disease conditions or to affect the structure or function of the body. This information is for educational purpose only. It is not to be considered medical advice, diagnostic or prescriptive.

To FMC Corporation's knowledge hydrogen peroxide has no proven therapeutic value for arthritis, HIV, cancer or any other ailments and does not promote supplying this material for these uses.

*** Hydrogen Peroxide H_2O_2 Secrets. THEY don't Want You to know.

Available at : http://www.h2o2-4u.com

(Accessed 6 June 2009).

Can't you see the miracle of water itself?

On the other hand, let me share you my believe about the three miraculous factors that make us growing well, staying alive & enjoying an abundant health as follow:- FOOD, WATER & HEAT.

- ***The soil that we cultivate, gives us food to grow well.***

  Foods are our body's vitamins.

- ***The water we drink, gives us life. Without water, we will never be able to survive, nor even all the living beings. Water composes of hydrogen combined with oxygen.***

  Water keeps us staying alive.

- ***The source of the heat we get comes either from the fire, or the sun. *** Dr. Hobday begins with the quite remarkable statement that there is a considerable body of scientific evidence demonstrating that? Sunlight may play a key role in preventing and ameliorating a number of serious degenerative and infection diseases, including cancers of the breast, colon, ovaries and prostate; diabetes; high blood pressure; heart disease; multiple sclerosis; osteoporosis; psoriasis; rickets and tuberculosis?***

  He describes the components of sunlight? Visible light, ultraviolet radiation and infra-red radiation, and the two wavelengths which affect the skin? Ultraviolet A (UVA? 320-400nm) and ultraviolet B (UVB? 290-320nm). Repeated exposure to the sun over many years leads to premature ageing, atrophy of the skin and even skin cancer.

*** The healing sun? Sunlight and Health in the 21st Century.

Available at:

http://www.positivehealth.com/Reviews/books/hob52.htm

(Accessed 21 May 2008).

However, heat gives us an abundant Health.

Since a direct exposure to fire may harm us, including the sun itself, if we get exposed to the sunlight for a long period of time, we may get harmed as well; however, the only factor that may give us an abundant health with no side effects or even causing any health complications is to heat water up to the boiling point (100° C), then wait for a while to cool at an affordable temperature we can afford (that is around 50° C), then gradually drink the HOT WATER. Nevertheless, we inhale enough quantity of Hydrogen while drinking hot water.

(Please refer to the benefits of Hydrogen as previously acknowledged in this book).

I strongly advise everybody who abides with the Miracle & Wonders of Treatment from Hot Water and its marvelous effects in our healthy life in particular, and in our whole life in general even for those who do not suffer any health complication, the benefits from drinking hot water is miraculous to never exceed more than two to three hours without drinking at least a glass of hot water. With this method, you are helping your body to keep your immune system strong in order to fight those harmful bacteria that exist everywhere. Mostly, our modern life has become much more critical. Despite the modern technology that has been tremendously been achieved, the side effects have increased in proportion with the technology. Let us take some few examples as follow:-

- *The air-conditioning emits billions of small particles in forms of bacteria from one part, and the gas being used to provide the cooling system, cause a lot of side effects that are the main causes of arthritis. Hot water helps you destroying such harmful bacteria and keeps your bones stronger enough to fight any form of arthritis, even with the elder people. Remember the wonders of hydrogen in our bodies, as previously mentioned.*
- *The chemicals that are used to preserve our foods, have played a big role to prevent a famine disaster that could occur, due to the huge increase of the world population not in proportion with the food supplies from one side, and the destruction of many farm lands and forests to build cities to accommodate such population increase in the other side. ***As of 27 september2009, the Earth's population is estimated by the United States Census Bureau to be 6.787 billions. World births have leveled off at about 134 million per year. Where as the deaths are only around 57 million per year. Here, is a simplified estimated world population in proportion with the times:-*

Year	*World (millions)*	*Africa (%)*	*Asia (%)*	*Europe (%)*	*Latin America (%)*	*Northern America (%)*	*Oceania (%)*
10,000 BC	*1*						
5,000 BC	*15*						
1750	*791*	*13.4*	*63.5*	*20.6*	*2.0*	*0.3*	*0.3*
1850	*1,262*	*8.8*	*64.1*	*21.9*	*3.0*	*2.1*	*0.2*
1950	*2,519*	*8.8*	*55.6*	*21.7*	*6.6*	*6.8*	*0.5*
1999-2000	*6,070*	*12.8*	*60.8*	*12.2*	*8.5*	*5.1*	*0.5*
2008	*6,707*	*14.5*	*60.4*	*10.9*	*8.6*	*5.0*	*0.5*

By the year 2000, there were 10 times as many people on Earth as there were 300 years ago.

***World population - Wikipedia, the free encyclopedia. (Updated 25 September 2009)

Available at : http://en.wikipedia.org/wiki/World_population

(Accessed 27 September 2009).

In return, such chemicals which are considered as toxins are definitely harmful and are the main causes of so many health complications and diseases.

- ***Most of the modern electronic and electrical equipments emit strong radiations, some very powerful. Such as microwaves, cell phones, computers, televisions, high tension electric lines...etc. They all made our life easier and more comfortable. But in return, such radiations have proven to be very harmful to our bodies causing many health complications. To the extent radiations are the main causes of various cancer diseases. Hot water is the best source to fight against any radiations. Remember what has been previously revealed in this book from some web sites about the effects of hydrogen in our bodies, that it is the ultimate antioxidant. And that cancer cells have no hydrogen in them at all. Simply by abiding the miracle & wonders of treatment from hot water, you will be able to inhale a large quantity of hydrogen that is coming out from the heated water in a form of steam. You will therefore be able to protect your body against any such radiations. But, prevention is***

better than cure. Help yourself by being vigilant and cautious from those harmful human made elements. We cannot by any means get completely rid of them, but we can surely have control by understanding their advantages and their disadvantages.

Remember as well, there is no any side effect or even any form of health complications in drinking hot water, as water is the only natural colorless and odorless element that no any living being can ever live without it. Even all the non-livings that serve all the living beings will never live without water. For example, our soil which is the main source of producing agricultural products for the foods we eat will completely dry without water. Water is the real source of life. Where as all types of man made medicines (whether traditional or modern), are flushed with toxins that they contain which our bodies completely reject.

Scientists have exerted tremendous efforts in various fields that have built the human civilization; they have also exerted huge efforts trying to find a complete cure for all diseases. Look at the tremendous technology that has been achieved by the scientists in making sophisticated medical equipments, capable of diagnosing the complete human's body with even very tiny strange objects within our bodies can now easily be detected. But, the technology in human anatomy for making human completely asymptomatic away from all health complications, remained a mystery until our modern world of today where we live.

Finally, let me share with you my recent discovery from "The Miracle & Wonders of treatment from Hot Water" as follow:-

By drinking hot water, by abiding the methods of drinking hot water as revealed in this book, without any interruption in a longer period of time, by completely believing in the miracle that you will be touched every time you drink a glass of hot water, your bones will become harder and elastic that will prevent you from having Osteoporosis (*a disease of bone, which leads to an increased risk of fracture. In osteoporosis, the bone mineral density -BMD- is reduced, bone microarchitecture is disrupted, and the amount of variety of non-collagenous proteins in bone is altered).**

*** Wikipedia, the free encyclopedia. Human skeleton

(Updated 24 September 2009)

Available at : http://en.wikipedia.org/wiki/Human_skeleton

(Accessed 27 September 2009).

You will also experience a very fast heal of any fracture in the same manner of a child bones, that may happen to you or a person who drinks plenty of hot water.

*** **A child's bones heal faster than an adult's because a thicker, stronger, and more active dense fibrous membrane (periosteum) covers the surface of their**

bones. The periosteum has blood vessels that supply oxygen and nutrients to the bone cells. The stronger and thicker periosteum in children causes a better supply of oxygen and nutrients to the bones, and this helps in the remodeling of the fractured bones by supplying. The periosteum in children causes a more rapid union of fractured bones and an increased potential for remodeling. A child's fractures not only heal more quickly, but are significantly reduced due to the thickness and strength of a child's periosteum.

*** Wikipedia, the free encyclopedia. Child bone fracture

Available at : http://en.wikipedia.org/wiki/Child_bone_fracture

(Accessed 27 September 2009).

Therefore, hot water which is the mix of oxygen and hydrogen, and by simply the water being hot it splits the hydrogen from the oxygen, will easily penetrate into our bones to perform such wonderful operations as aforesaid.

As previously mentioned in this book from some websites, HYDROGEN IS THE BODY'S MOST NEEDED NUTRIENT.

That is why babies rarely experience bone fractures when falling down, including some insects and animals such as the lizard, snake…etc. compared to adult people!!!

Isn't this a miracle? Truly it is.

With all that has been provided, I would like to share you my believe that "**OUR BODIES NEVER ACCEPT STRANGERS, WHETHER MODERN OR TRADITIONAL MEDICINES, OR ANY FORM OF HUMAN MADE ELEMENTS, SUCH AS TOXINS, RADIATIONS... etc**"

HOWEVER, YOU ARE STRONGLY ADVISED TO BELIEVE ABOUT THE MIRACLE & WONDERS OF HOT WATER.

Finally, are all the information that I have been able to provide you from my various personal researches, experimentations and those that I have collected from different sources, profitable to you?

If no, I would like to tender my strong apology. And promise to do all my best to provide you with more fruitful information about the miracle & wonders of treatment from hot water, in my second edition of this book.

If yes, then your recognition and believe is considered an award certificate to me in being able to achieve a true leadership which I would like to share you my theories as follow:-

"**A true leadership is the one leading others achieving their ambitions, solving their issues, removing their obstacles, contributing in the building and progress of the Human Civilizations, making real peace and achieving stability in all the fields**".

Moreover, "**One of the miracles of life is that, every Human being has been born as a leader. The star** leading him to enlighten his leadership, arise in proportion of his surrounding environment. Some achieve a dynamic leadership leading them to a sustainable success. Others achieve a wiggle leadership, leading them between success and failure. Whereas others as well, never intent to enlighten their leadership, leading them to a complete failure in their life. However, the only secret to accomplish a sustainable dynamic leadership depends upon our faithfulness to face all the challenges confronting us peacefully and accepting whatever the results arising thereafter".

Written by;

Faris Rashid Salim Al Hajri
Born in Africa, State of Burundi, October 27, 1964
Married to a Nurse, Omani nationality, originally Philippino.
Father of two Boys, sharing the same birthday, October 28, 1996 & October 28, 2002.
Qualification: High level Diploma in Quantity Surveying, Construction Engineering from Oman Technical College 1987, Sultanate of Oman.
Actual Position: Technical Expert in the Office of H.E. the Undersecretary of the Ministry of Housing Sultanate of Oman

October 3, 2009

Note:-

* *The result period may differ from one Person to another.*

Theory # : *Serial No. of my Theories* about "THE VALUES OF WELL BEING & ITS SECRETS, FOR A BETTER LIVING"

4- *Note:- Everything that has been noted in this book, has been revealed upon my personal efforts and various experimentations I went through for many years in one part, and many researches and information that I have been able to collect related to the topic itself in order to increase the credibility of the contents mentioned in this book, in the other part.*

Drinking of hot water has been a wonderful scientific discovery I have ever accomplished. I have myself abided in drinking hot water every day of my life that has now lasted for more than two and half years, the moment I discovered this method.

Whereas I still continue drinking hot water and will never boycott it, due to its great results I have accomplished.

Personally, I was able to cure myself from so many complications and diseases, some of them were considered chronic to the extent they were incurable with either the modern or traditional medicine.

Moreover, a tremendous number of People have benefited, and many of them are now completely asymptomatic from their diseases. Therefore, I wish that every Human being will benefit from the miracle and wonders of treatment from hot water, including every one who read and abide by what has been revealed in this book.

In addition, simply since water is the only element that no living being can ever live without it, and is the only colorless, odorless and tasteless element that never cause any harm to any living being including Human himself, therefore, I do not bear any responsibility toward any Person if wrongly use this discovery. As the Person who drinks hot water and follows the procedures mentioned in this book, is considered a doctor for himself, a volunteer by himself and seeks for the benefits of the miracle and wonders of treatment from hot water, for which it is my wish to every Human being.

This note is not pertained to be neither a warning, nor an apprehensive from the treatment. But, for the sake of completing the legal procedures of this book, by preventing any wrong intention that may arise from any Person for the sake of achieving his personal needs that are completely denied and do not constitute any part of the goals for which this book has been issued.

Made in the USA
Columbia, SC
23 October 2018